CURIOUS CREATURES

LIFE IN
THE DARK

Written by
Joyce Pope

Illustrated by
Stella Stilwell and Helen Ward

STECK-VAUGHN
L I B R A R Y
A Division of Steck-Vaughn Company
Austin, Texas

Editor: Andy Charman
Designer: Mike Jolley
Picture research: Jenny Faithful

Library of Congress Cataloging-in-Publication Data

Pope, Joyce.
Life in the dark / written by Joyce Pope.
p. cm. – (Curious creatures)
Includes index.
Summary: Describes the habitats and activities of
creatures that live in darkness.
ISBN 0-8114-3150-9
1. Nocturnal animals – Juvenile literature. 2. Deep-sea fauna – Juvenile
literature. 3. Cave fauna – Juvenile literature.
[1. Nocturnal animals.] I. Title. II. Series.
QL755.5.P67   1992   91-18646
591–dc20   CIP AC

NOTES TO READER
There are some words in this book that are printed in **bold** type.
A brief explanation of these words is given in the glossary on p. 45.

All living things are given a Latin name when first classified by a scientist.
Some of them also have a common name. For example, the common name
of *Sturnus vulgaris* is common starling. In this book we use other Latin words,
such as larva and pupa. We make these words plural by adding an "e",
for example, one larva becomes many larvae (pronounced lar-vee).

Color separations by Positive Colour Ltd., Maldon, Essex, Great Britain
Printed and bound by L.E.G.O., Vicenza, Italy

1 2 3 4 5 6 7 8 9 0 LE 96 95 94 93 92

CONTENTS

LITTLE OWL

LEOPARD CAT

VAMPIRE BAT

LIFE IN THE DARK

Human beings are creatures who need light. In early times, people lengthened the daytime with firelight, candles, and lanterns. Now, our towns and cities are lit mainly by electricity and there are some places where it never seems to be night. Even today, some people are afraid of the dark. This is probably because our eyes are the most important of our **senses**, and eyes work only where there is light. We feel lost and in danger when our eyes cannot tell us something about our surroundings.

There are many parts of the world where it is always dark. The depths of the oceans, caves, and the polar regions in winter are all places that the sun never reaches. Animals live in all of these places. Some make their own light. They often have huge eyes, which can see even small amounts of light. Others have no eyes and rely on other senses, particularly their hearing, and their senses of smell and touch.

If you were to see a completely unknown kind of creature, you could probably make a good guess as to whether or not it lived in the dark. This is because creatures that live in the dark have certain features that help them to survive. This book is all about creatures that live in darkness.

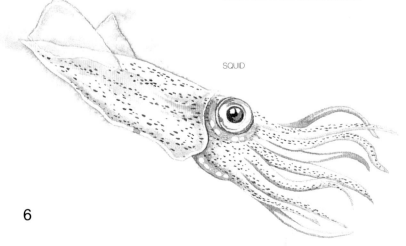

SQUID

◀ ▶ Here is a collection of creatures that live in the dark. Twilight animals, such as owls, have huge eyes; so do deep-sea fish and squids that make their own light. The slender loris has big eyes and good hearing to help it spot **predators** or find food in the dark. Bats find their way about by using their hearing. Cats have big whiskers to help them feel their way.

◀ The giant water lily lives in slow-flowing streams in the rain forests of Brazil. Its cabbage-sized flowers attract many kinds of insects, but particularly cockchafer beetles. When the sun goes down, the petals close. The beetles cannot escape, but they feed on the flower's stamens, and get a good deal of pollen on themselves. Next morning, the petals open again. The beetles fly off, carrying pollen to other water lily flowers. These flowers are then **pollinated**. As a result of this, the big spiny fruit of the giant water lily is formed.

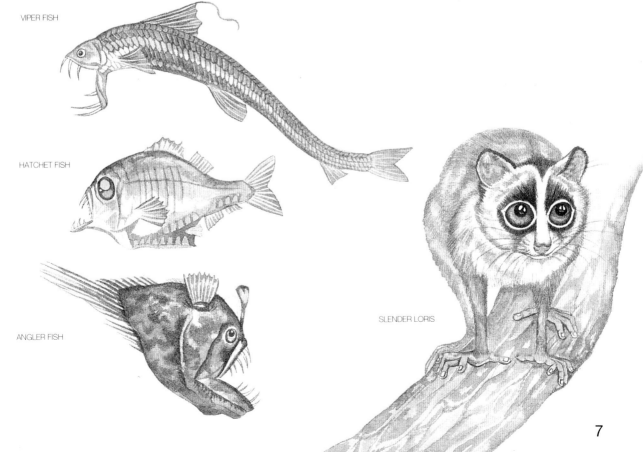

VIPER FISH

HATCHET FISH

ANGLER FISH

SLENDER LORIS

7

LIFE AFTER THE SUN

DUKE OF BURGUNDY FRITILLARY

When the sun goes down at the end of each day, three things happen. The first, and most obvious, is that it becomes dark. The next is that it becomes cooler. The third is that the air becomes more moist and dew forms as the temperature drops.

NOCTURNAL LIFE

Coolness, darkness, and moisture do not suit all kinds of creatures. Many creatures that are active in the daytime, hide and sleep at night. Other creatures come out of their dens to feed and find their mates in the dark. They are called nocturnal animals.

The night sky is rarely completely dark. Often the moon and stars give some light. Some nocturnal animals, like owls and night monkeys, have large eyes

▲ The moth and the butterfly shown here look very similar, but there are several important differences. Moths, which fly after dark, usually have dark colors and thick fur to help keep them warm. Butterflies, which are active in the sunlight, are far less hairy.

▼ On most seashores the state of the tide is more important than whether it is day or night. Birds such as the oystercatchers can only feed when the tide is out. The limpets move and feed only when they are covered with water. As a result of this, there is always some activity at night on the seashore, but it is not always the same kinds of animals.

that are able to use every bit of light.

Other nighttime animals use their hearing. Many, such as cats, are good at knowing the direction from which a noise comes. Some, such as bats, use **echoes** of sounds to avoid bumping into things and also to catch their food. Other types of animals have sensitive whiskers around their snouts. When these touch anything, they warn the animal of an obstacle.

▲ As the sun goes down, the brightly colored animals that are active during the daytime find somewhere safe to rest. Their place is taken by animals that are mainly gray and brown. These nighttime creatures keep in touch with each other by making sounds or using their sense of smell, because as it gets dark it is more difficult to see colors.

THE NIGHT SHIFT

Almost all energy on earth comes from the sun. You may wonder, then, why any animals should be active at night. The main reason is that the environment is like a factory. It has a bigger output if it works all the time. Nature has a day shift and a night shift. Even plants, which can only grow in the light, may have flowers that open in the dark. These feed creatures of the night shift.

Often, you can find two kinds of animals that live in the same manner, except that one is active in the daytime and the other is active at night. Butterflies are creatures of the sunlight. Their cousins, the moths, almost all fly at night. Both use their long tongues to feed on the **nectar** produced by flowers.

Many small insects fly in daylight and are hunted by birds such as swallows, swifts, and flycatchers. After dark, bats take the place of the birds. Bats eat mainly night-flying midges, beetles, and moths.

Hawks and eagles hunt in the daytime, owls hunt their **prey** at night. These two kinds of hunters both have sharp, hooked beaks, and talons. They are not closely related though they look alike.

▶ Night-flying insects like this moth feed mainly on nectar that they take from flowers. Sometimes soft-skinned fruits, like figs or plums, crack when they are ripe. The sweet juice that flows out gives moths and other night-flying insects the sugars they need for food.

▼ Below left you can see the edge of a forest as it looks by day. Below right shows how it looks at twilight. By day, butterflies and bees gather nectar from flowers. At night, the strong scent of the honeysuckle attracts moths. Moths are hunted by bats, which come out at night. A hawk may catch a squirrel by day; owls and foxes come out to hunt at night.

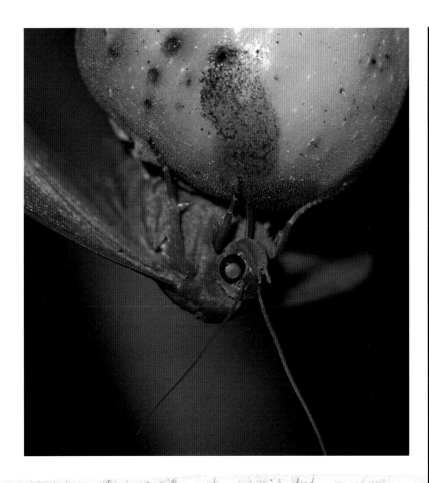

Some nocturnal animals, such as the insects shown below, seem to be attracted to light. They often blunder into brightly lit rooms, where they are dazzled and confused. The reason for this is that they are able to find their way using the moon's light. They fly so that it strikes part of their eyes. They keep going in a straight line because the moon is so far away. When candles and electric lights are nearby, the insects treat them in exactly the same way, and end up flying in circles.

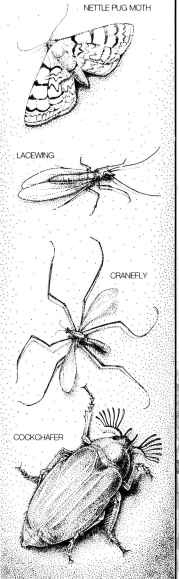

NETTLE PUG MOTH

LACEWING

CRANEFLY

COCKCHAFER

SAFETY IN THE DARK

For many small and helpless animals the darkness means safety from enemies. Animals such as mice wait until dark before they come out of their holes. Nighttime hunters, including owls, foxes, and cats, catch many of them. But the mice would be in even more danger if they came out during the day.

The coolness and moisture of the night suit other animals. Worms leave their tunnels to search for leaves. Worms avoid light; their moist bodies would dry up in the heat of the day. Slugs and snails also come out at night to feed on soft or rotting vegetation. Although they move slowly, they are quick to spot danger. Worms shoot back into their tunnels, snails go back into their shells, and slugs rely on their bitter slime to deter predators. These tactics are not always enough. Badgers dig up and eat enormous numbers of worms. One badger may eat several hundred earthworms in a night. Hedgehogs and shrews eat snails, and small slugs are the favorite food of slowworms.

In warm places, many **cold-blooded** animals are active after dark, because the days are too hot for them. **Reptiles** and **amphibians** find the small prey that they need among the insects and **rodents** that live near them.

▲ Many small animals live in gardens like the one shown here. On summer nights, worms come to the surface of lawns, and snails and slugs feed on soft plants. They are hunted by creatures such as beetles, toads, raccoons, and opossums that usually stay hidden during the daytime.

▶ Small birds, like these swallows, often migrate at night and feed during the day. In daylight hours, big flocks of migrating birds are at risk from hawks. In the dark, they usually fly above the height at which they would be in danger from owls.

Taking Shelter

Shearwaters spend most of their life at sea. They only come ashore on remote breeding grounds, where they lay their eggs in the safety of a burrow. They go to and from their burrows at night. Even bright moonlight can mean danger to them because, although they are very good fliers, they are nearly helpless on land. Large gulls patrol the shearwaters' burrows and attack any birds returning late from feeding trips.

Don't Badger Us!

Badgers are active at night, though they may come out to feed during the daytime if they are not disturbed. European badgers are probably the most nocturnal. Badgers have become very wary and, over time, they have developed the habit of avoiding people whenever possible. Like badgers, many other animals remain in hiding until after dark, when they are least likely to meet their human enemies. Nowadays, many naturalists enjoy badger-watching, but they have to do it at a time that suits the badgers.

THE DESERT AT NIGHT

Animals have other enemies in addition to predators. In many parts of the world the climate is an enemy. This is especially so in hot deserts, where daytime air temperatures may go up to more than 130°F. Very few living things can stand such intense heat. For this reason, there are very few animals active during the day in deserts, but many that appear after dark.

More than one-fifth of the world's land surface is covered by desert. This is an area the size of Africa. Hot deserts are found in parts of North and South America, Africa, Asia, and Australia. Although the surface looks dead, many kinds of creatures live beneath it. The heat does not reach them and the air in their burrows has more moisture than the air above.

As darkness falls the temperature drops and the desert comes to life. Most kinds of desert **mammals** have large ears which allow them to tell the direction of any sound in the dark. The rattle of a pebble or a tumble of sand grains may mean a meal or an approaching predator. Their huge ears also have another function. They help the animals to control their body heat and keep cool.

▼ Deserts have fewer animals than areas with more water. You can see some desert-dwelling animals below. As in all places, these animals depend on plants for their food. Plants supply food for insects that are hunted in turn by other insects, scorpions, reptiles, and small mammals. In the driest parts of deserts, plants appear after rain. The number of small animals rises at the same time. The plants flower quickly and soon die, but their seeds remain. These are collected and stored for food by small rodents, such as jerboas or kangaroo rats, which can survive their entire lives without drinking. These small rodents are eaten by predators that survive on liquids from the bodies of their prey.

▲ During the heat of the day the jerboa rests in a burrow several feet below ground. After dark it emerges to search for the seeds and roots that are its food. Jerboas leap like miniature kangaroos. The broad, hairy toes on their long hind feet help them not to sink in loose and shifting sand. Jerboas live in the deserts of north Africa and Asia. There are similar rodents in other parts of the world.

Nighttime Pollinators

Some desert plants have special pollinators. The saguaro cactus of the North American deserts is visited at night by long-nosed bats. Pollen gets stuck to the bats' fur while they are feeding and is then carried from one flower to another. Another nighttime pollinator of the American deserts is the yucca moth. A female moth pollinates a flower with a carefully carried ball of pollen. She then lays eggs on the plant. These hatch into caterpillars that feed on some of the seeds the plant makes. In Australia, some plants are pollinated by small mammals. They may be searching for insects, but they pollinate the flowers as they do so.

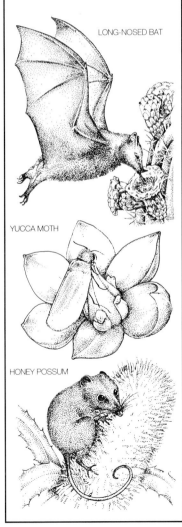

LONG-NOSED BAT

YUCCA MOTH

HONEY POSSUM

DARK PLACES

Caves and the deep ocean are the darkest places in the world. In the winter, sunlight does not reach the areas around the North and South Poles. They are dark for a whole season. In other places some burrowing animals never come to the surface and never see light.

HUNTERS IN THE DARK

Plants cannot live where it is always dark. This is a problem for animals that live deep in caves or in the depths of the ocean. The basis for all their food must

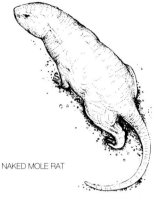

NAKED MOLE RAT

▲ Naked mole rats spend all of their lives underground. Groups of at least 20 animals, led by a large female, live together in desert areas in east Africa. Generally the young ones dig burrows, which may total well over 300 yards. They also find the roots and tubers to feed the colony.

► Many reptiles live in burrows, but the worm-lizards, shown right, spend all of their lives underground. Worm-lizards are blind and, in almost all cases, legless. They use their heavy heads to dig. Their scales hold their position as they tunnel, and the lizard can move forward and backward equally easily.

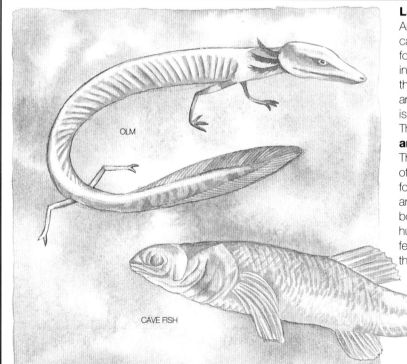

OLM

CAVE FISH

Living in a Cave

Animals such as bats take refuge in caves, but rely on the outside world for their food. Other animals, including those shown here, spend their whole lives in the dark. Most are blind and white – **camouflage** is useless where there is no light. These animals usually have long **antennae** or feelers on their bodies. These sense the slightest movement of air or water, which may mean food is nearby. A few of these animals are scavengers and eat the bodies of dead animals. Most are hunters. They are among the most ferocious of all animals in spite of their small size.

▶ In the Antarctic, where most mammals depend on the ocean, these emperor penguins are the only animals that spend the winter on land. The penguins start their long breeding season at the beginning of the most bitter weather. Each male guards a single egg in darkness and cold for over two months without leaving to get food. The penguins huddle together to keep warm.

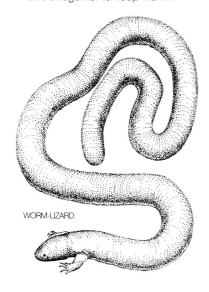

WORM-LIZARD

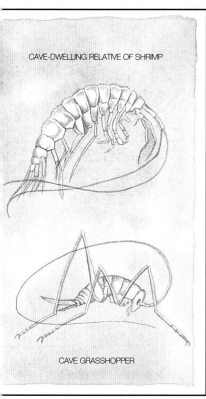

CAVE-DWELLING RELATIVE OF SHRIMP

CAVE GRASSHOPPER

come from parts of the environment where there is plenty of light. Plant fragments drift down from the surface of the ocean, or are washed into caves by rivers.

Eyes are of no use where there is no light. Most cave animals are completely blind and rely on their senses of touch and smell to tell them about their surroundings.

Some deep-sea animals are blind. Others have huge eyes and go into areas where there is a little light. Certain animals make their own light.

Some burrowing animals rarely come to the surface. Like moles, these animals usually are blind or nearly blind, though they generally have a good sense of touch, smell, and hearing. Some insect grubs eat their way into wood. They are among the most helpless of all animals, but they are surrounded by their food, which protects them from most predators.

These creatures are the basic food for a number of different kinds of flesh-eaters, or **carnivores**. These flesh-eaters are usually small and rare, but they are very good hunters.

Squid of the Deep
Sandalops has been found in water about 3,000 feet deep. Like many creatures of this cold, lightless zone, it is tiny. The length of its baglike body is only half an inch, so the drawing here is larger than life-size. Its eyes pick up any trace of light made by other animals. It also has a pair of light-detecting organs on its underside. It is thought that these may be sensitive to the light that *Sandalops* produces itself and so enable it to compare its own light with that made by other animals, which may be food or enemies.

▶ This prawn lives in the deep sea, below about 2,500 feet. Although it looks so bright here, in the blackness of its own habitat, where there is no red light, it would be quite invisible. Its long, sensitive antennae enable it to detect water movements and avoid enemies or catch a meal.

THE DEEP OCEAN

The oceans cover about two-thirds of the earth's surface. Their average depth is about 12,000 feet. The deep ocean is, therefore, the largest of all the world's habitats. In spite of this, it is the habitat with the smallest number of creatures. This is because most of the world's seas and oceans are dark. The sun's rays reach down only a few feet. The animals that live there lead a life without seasons. In the deep ocean there is no change in light or warmth that may be taking place above the waves.

Most of the oceans' life is in the well-lit surface waters. Dead plants and animals float from these surface waters toward the ocean floor. There, they form the start of a **food chain** for dwellers in the darkness. As in caves, there are many ferocious carnivores in the deep parts of the ocean. The fish have huge teeth and many can swallow a meal larger than themselves.

The darkness of the deep oceans is broken by light that is made by some of the animals. Many kinds of fish and other animals are able to produce their own light. This light may be used to keep in touch with others of their own kind and to attract the opposite sex. Light may also be used to trap food or to frighten predators.

▼ ► Even if you dive in the clearest seawater, you will find that it begins to get dark very quickly. All red and orange light disappears before you reach a depth of 90 feet. Green and blue light lasts longer, but at a depth of 600 feet even it is gone. Below this is complete darkness. All living plants and most animals are in the top few feet. Some surface-living fish and whales may dive briefly into deeper water, but do not stay there long. In the deeper parts, which run down to over 30,000 feet, some of the strangest of all fish and squids live. If you could dive into the cold, dark waters of the deep sea you would see some of the animals shown here.

BLUE WHALE

GREAT SWALLOWER

SQUID

LANTERN FISH

VIPER FISH

HATCHET FISH

ANGLER FISH

TRIPOD FISH

SURVIVAL KITS

▶ Most kinds of small cats, such as this jungle cat from Southeast Asia, are nocturnal. They hunt for mice and other small prey which they find by listening for the slight sounds of their movements. The cats themselves move almost silently. They can pull back, or retract, their claws to make their feet soft and silent. In spite of this, the power and sharpness of the claws enables them to kill quickly.

Animals that feed or hunt in the dark need a survival kit of sharp senses. Sight, hearing, smell, and touch are all used in the dangerous task of finding, but not being, a meal. Some animals have special senses, such as the ability to detect another creature by the heat of its body.

▶ Many nocturnal animals such as this spiny anteater and the tenrec, are protected by a spiny coat. Porcupines also have spines. Even large carnivores stay away from porcupines because their spines work their way into the paws or noses of attackers, often causing serious injury.

SPINY ANTEATER

TENREC

NINE-BANDED ARMADILLO

◀ This nine-banded armadillo usually hunts at night. Protected by its bony shell, it wanders about in search of small prey such as lizards, frogs, beetles, and woodlice. Few enemies frighten the armadillo. It even ignores human beings. Perhaps it does not bother to be quiet because it is so well-armored. It grunts almost all the time as it roots about.

SHARP EARS AND CLAWS

Nighttime hunters need to find their prey, and to catch and kill it quickly. Many track their prey by using their sense of smell. Others hear their prey, or, under water, feel that it is there. These senses of hearing and touch give hunters an idea of where another animal may be hiding. They are then able to creep up on it, moving so quietly that they are able to pounce at the last moment.

Animals that are hunted are also silent. Only animals that are very well-armored are safe making noise. The hedgehog, which is protected by its prickles, sounds like a much bigger creature as it pushes through dead leaves. Shrews squeak as they bustle about hunting smaller animals. They have powerful stink **glands** and other mammals will not normally eat them. They are caught by owls, which have no sense of smell and can tell exactly where shrews are because they make so much noise.

Nighttime Hunters

Cats and owls are both night-time hunters. Both have the same need for silence. A cat's feet can move noiselessly; an owl's wings have soft feathers, so that its flight is silent. Both cats and owls use their eyes when hunting. The **pupils** of their eyes close to a very small area in the daytime. At night, they open wide to catch all the light available. Both use their ears to find the exact position of their prey. Cats can turn their ears to the direction of any noise and so can some owls. Cats kill with their claws and slice with their teeth. Owls also have claws, or talons, and a narrow, sharp beak for tearing food.

OWL EYES

CAT EYES

OWL TALONS

CAT CLAWS

POISON IN THE NIGHT

In warm countries, the night shift includes many creatures that are small or slow-moving. Some of these animals use poison to escape from predators or to catch their prey. Often their poison, or **venom**, is extremely powerful, but the amount that they produce is so small that they are quite harmless to human beings. Also, most of these animals are shy, and avoid people whenever they can. The danger comes when they are surprised and cannot escape.

Venomous animals of the night find their prey in many ways. Some spiders spread webs to trap night-flying insects. In a similar way, the **tentacles** of corals active at night catch tiny animals swept to them by the sea currents. Scorpions have long hairs on their pincers. These sense tiny movements in the air, such as those that would be made by small animals. The rattlesnakes and their relatives are called pit vipers. They have a pit

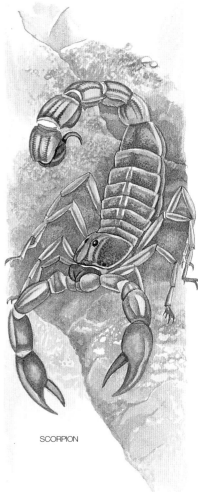

SCORPION

▲ Most scorpions, like the one shown here, spend the heat of the day in burrows or under stones. They do not come out until the evening, when it is cooler. Their main prey is large insects, which they sieze with their pincers. The prey is **paralyzed** by an injection of poison.

◄ This frog is protected from predators by poison glands set in its skin. Toads and salamanders are also protected in this way. If a predator seizes one of these animals, the glands produce a bitter liquid. This makes the hunter, which is usually unharmed, drop its prey.

▼ A sidewinder swallowing its prey. Pit vipers such as this are some of the most venomous of all snakes. In spite of this, they usually catch small animals. They find their prey at night with the aid of sensitive areas on either side of the face called pits. These pits can detect the heat given off by another animal's body.

SIDEWINDER

just below their eyes. This works as a heat-seeking organ. It enables them to build up a picture of the heat pattern of other living things. They can strike at a mouse, even in complete darkness, thanks to this picture.

In almost all poisonous animals, the venom works in two ways. First, it paralyzes the prey so that it cannot move or escape. Then, part of the poison acts like a digestive juice, and begins to break down the food. Many venomous creatures cannot chew their food, but the venom helps them to digest it quickly.

Web-throwing Spider
Web-throwing spiders are sometimes known as ogre-faced spiders, for they have two huge eyes, that some people think make them look fierce. During the daytime, these spiders, which are camouflaged to look like twigs, hide in bushes. At night, they make a small, very sticky web of extremely strong, elastic silk. The spider hangs by its hind legs waiting for an insect to pass by. As it does so, the spider stretches its legs and opens the net. The insect becomes entangled and is eaten through the web.

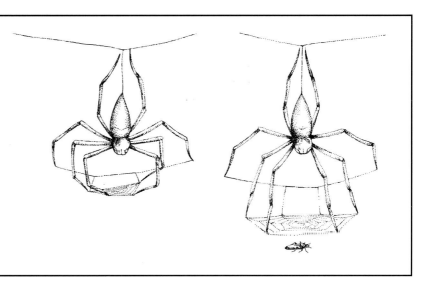

SEEING IN THE DARK

Eyes are like cameras set into our heads. Like cameras, they have lenses through which light passes, and a light-sensitive background on which the picture of our surroundings is formed. Most animals that live in bright light, such as birds, many kinds of insects, some reptiles, and fish that live in shallow water, can see colors. In fact, some insects can see colors that humans cannot. Animals that live in twilight or dark conditions often have eyes that are very sensitive to light, but they cannot see colors. Animals that live in complete darkness are often blind, because eyes, like cameras, can only work where there is some light.

Animals with backbones, such as birds and mammals, have eyes with pupils. Light enters the eye through the pupil. The size of the pupil changes. In very bright light it is closed by the colored **iris** to a narrow slit or pinpoint. When the light is dim, the pupil opens wider, so that the

▼ All of these small mammals live in tropical forests. They hide in hollow trees and other safe places during the daytime. At night, they come out to hunt for insects and other small creatures. They all have very large eyes, which can use every scrap of light. In spite of this, they are almost certainly color-blind. This may be one of the reasons that they are all dull colors themselves.

POTTO

LESSER BUSHBABY

GREATER BUSHBABY

RUSSET
MOUSE LEMU

Eyes Reflecting Light

The eyes of some nocturnal animals reflect a greenish glow in a beam of light such as a flashlight or a car's headlights. This is because they have a reflective layer, called the tapetum, in the retina. It is thought that the tapetum enables the animals to re-use light that has already entered their eyes. Some animals have eyes that glow red in a beam of light. These do not have a tapetum. The redness is a reflection of blood vessels over the retina. Many kinds of night-flying insects, as well as shrimp and crabs that live in the dark, also have reflective layers in their eyes.

iris almost disappears. In each case the light is focused through the lens onto the sensitive **retina** at the back of the eyeball. The retina is made up of two different types of **cells**. Cone cells give color vision. Rod cells contain a **pigment** that enables them to detect light. Animals such as owls or deep-sea fish that live in poor light have huge numbers of rod cells. As a result, they can see quite well in conditions that seem completely dark to humans.

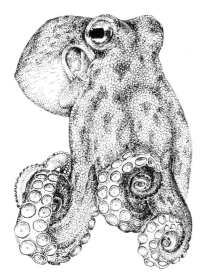

OCTOPUS

▼ Octopuses and their relatives the squids and cuttlefishes can see well, for their eyes are very similar to those of mammals. Like many animals with all-round vision, the octopus has an oblong pupil. The pupil of the gecko's eye opens wide at night, but during the day closes to a narrow slit with a series of "pinholes" through which light passes. Scientists think that this gives the gecko very sharp vision, probably better than that of most other animals.

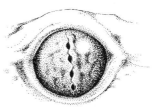

CLOSE-UP OF GECKO'S EYE

TOKAY GECKO

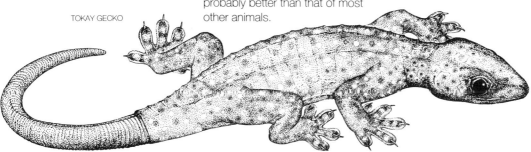

ALL EARS

The sense of hearing is important in places where there is little light. Animals that live in such places cannot ignore the sound of the crackle of a twig or the rattle of pebbles as something scuttles around. Being able to hear the sound is a matter of life and death.

The actual work of hearing happens inside the head of an animal. The large ears on the sides of mammals' heads are only part of the system. They can often be twitched in any direction to pick up sounds more clearly. Other nocturnal animals do not have these ear trumpets, but they usually have good hearing.

Not all sounds that are heard at night are accidental giveaways of an animal's position. Many creatures find their mates at night. Male frogs and toads croak to attract females. Some fish make drumming or rattling noises that can be deafening underwater.

Insects such as bush crickets and cicadas also make noises at night, and some birds, such as nightingales, continue their singing after dark. All of these animals are instantly quiet if they hear a sound that may mean danger.

TARSIER

BAT-EARED FOX

◄ ▼ Here, you can see a collection of animals that have a very good sense of hearing. Tarsiers and bat-eared foxes, with their huge ears, are alert to every sound. Frogs have eardrums on the side of their heads. Cicadas have similar eardrums on their sides. Bush crickets have their ears on their front knees. They can tell the direction from which a sound is coming by moving their legs.

EDIBLE FROG

DARK BUSH CRICKET

▲ The tiny fennec fox has the largest ears, compared to its size, of all flesh-eating mammals. Its diet includes many insects. It can leap up to 30 inches from the ground to catch them. Fennec foxes also dig to find creatures underground and may be able to hear movements that are made below the sand.

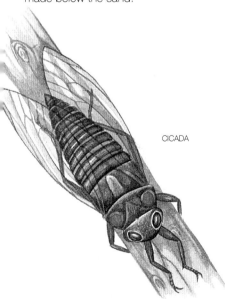

CICADA

BARN OWL

Hunting by Sound

Some mammals are able to judge the distance and direction of sounds by using their movable ears. This is much more difficult for creatures that do not have external ears. Some of these animals solve this problem with ears placed in slightly different positions on either side of the head. This means that sounds reach each ear at slightly different times. As a result, the animal's brain can work out the direction and distance of any sound. The barn owl shown here can hunt successfully in total darkness in this way.

ECHOLOCATION

Humans and other animals hear sounds when pressure waves in air or water strike their eardrums and make them **vibrate**. The vibrations are passed by tiny bones to the **inner ear**, where they are turned into electrical impulses that can be understood by the brain. The faster the vibrations, the higher the pitch of sound that they hear.

Sound waves can bounce off an obstruction – that is, anything that gets in the way. The sound waves then make a new noise, which is an echo. Sound waves travel at a known speed, so it is possible to work out the distance between the thing that is making the sound and the thing that caused the echo. This is called echolocation or sonar. In the 1930s it was discovered that bats use echolocation to find their way and catch their prey, even in complete darkness. It is now known that several kinds of animals use echolocation.

An animal using echolocation makes a series of squeaks or clicks of sound. These sounds are usually too high-pitched for human ears to hear. If the sound waves hit something, an echo will be made. When the animal hears this, it will take action to avoid or chase the object that is in the way.

Cave-dwelling Birds

South American oilbirds and several kinds of cave swiftlets, all of which roost in caves, find their way in the dark by echolocation. The sounds that they make are low-pitched clicks. None of the birds uses its sonar in the light. They do not use it to find or catch food because they all have good eyesight. The swiftlets are active during the daytime and feed on insects. Oilbirds leave the caves at night and eat fruits such as nutmegs. Some fruit-eating bats also use clicking noises. Click sonar does not give such accurate information as high-pitched squeaks.

▶ Bats are not blind, but they have small eyes and poor eyesight. The size of their ears shows us that hearing is their main sense. Bats' ears are large and often have a big extra lobe called the tragus, which probably helps to guide sounds into the ear. Bats make many sounds as well as the high-pitched squeaks that they use when hunting.

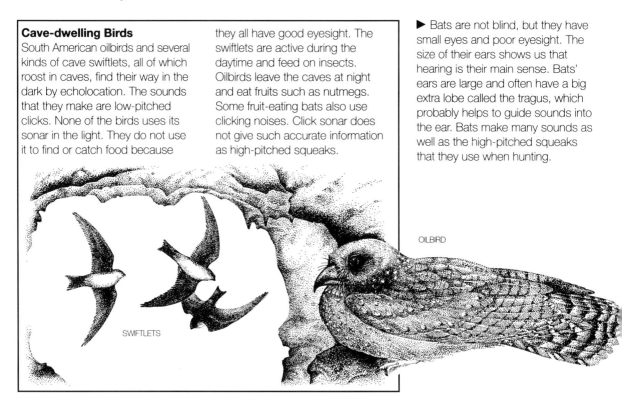

OILBIRD

SWIFTLETS

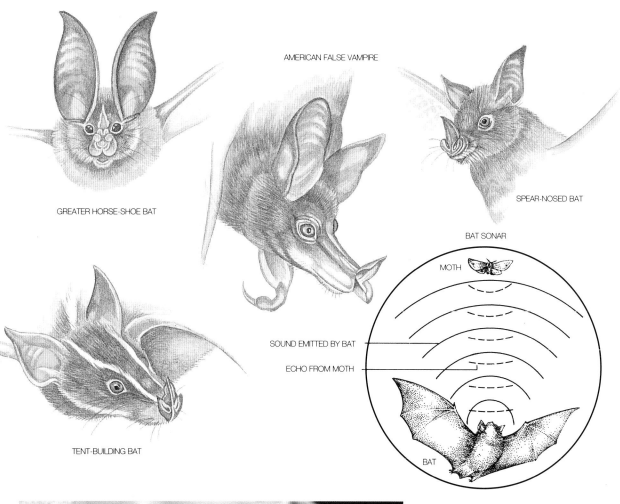

AMERICAN FALSE VAMPIRE

GREATER HORSE-SHOE BAT

SPEAR-NOSED BAT

BAT SONAR

MOTH

SOUND EMITTED BY BAT

ECHO FROM MOTH

BAT

TENT-BUILDING BAT

▲ Some kinds of bats have doglike faces. Many others are bizarre looking creatures, with flaps of skin around their noses and jaws. The purpose of these is almost certainly to focus the high-pitched squeaks that they make for echolocation. These are most effective if they form a narrow beam of sound, for in that way the bat will not be confused by echoes from objects on either side of it. As it flies, a bat makes four or five squeaks each second. When it hears an echo it increases the rate to about 200 bursts of sound each second. From the echoes, the bat can work out whether the obstruction is still or moving. Each bat recognizes its own voice, so even when several bats are flying together they do not get muddled. High-pitched sounds die away very quickly, so echoes from distant objects do not bother the bat.

SOUNDS OF THE DEEP

At one time the sea was called "the silent world." Fishermen knew that many strange noises came from the water, but most people thought that ocean-dwelling animals were silent. It was not until underwater microphones, called hydrophones, were first used that people realized how noisy the oceans are. Sounds travel farther in water than in air. It is possible that a humpback whale's song may be heard by others hundreds of miles away.

We do not know the reasons for all the sounds heard underwater. Some may be made by animals hoping to attract a mate. Others are used by groups of animals to help them keep together in the dark water. Different ones are alarm sounds and others are used for defense and attack.

Most animals that live in the ocean do not have voices. Whales and dolphins make whistling and clicking noises through their blowholes. Many fish make a drumming sound by twanging a special muscle against their **swimbladder**. Some animals without backbones use their armor to make rubbing or clapping sounds.

▼ Sperm whales, like the mother and calf shown here, make clicking sounds which travel well through the water. Each whale makes its clicks in a special pattern so that others in the group will know where it is. Sperm whales dive deeper than any other whales, often over 3,000 feet.

▼ Dolphins often hunt in groups, or schools. The school shown below is hunting mackerel. Dolphins find fish by echolocation and often drive and bewilder them with bursts of high-pitched sound. They have to be careful not to direct the sound at each other. It is dolphin "good manners" to stop the killer sounds when they are facing each other. Dolphin sounds are made in the lower part of their blowholes.

Killing with Sound

The pressure waves from bursts of sound made underwater may be strong enough to use as a weapon. Snapping shrimp make a cracking noise with their large claws. This noise disables their prey. Dolphins are able to concentrate the noise they make into a narrow beam.

People have seen them attack large fish with bursts of sound. The fish loses its balance and is easily caught. Healthy sperm whales have been found with oddly-shaped jaws. These whales would not have been able to catch their prey normally, so they probably killed with sound.

FEELING THEIR WAY

The sense of touch is very important for animals that live in the dark. By using antennae or long, sensitive whiskers. they can find out about their surroundings without disturbing things. This is important, because if they made a noise, or fell over things, they might be discovered and eaten by an enemy lurking nearby. If they were noisy they would also lose any chance of finding their own food.

Most mammals have large whiskers, called vibrissae, around their snouts and sometimes on other parts of their faces. There is a large **nerve ending** at the base of each hair, which enables it to sense the slightest touch. Moles have very sensitive tips on their noses. Star-nosed moles have 22 little fleshy tentacles on the end of their snouts. They are able to find grubs and worms in the dark with these tentacles.

Many animals can detect movement in the air. One reason that cockroaches are difficult to kill is that they can feel the rush of air caused by a rolled-up newspaper or similar weapon when someone tries to swat them.

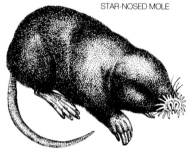

STAR-NOSED MOLE

▲ Star-nosed moles get their name from the 22 short, pink, fleshy tentacles around their nose. As they hunt for earthworms, insects, and freshwater shrimps, they hold the two rays immediately above their nostrils stiffly forward. The rest are constantly moved, feeling for prey. Baby star-nosed moles are able to search for food in this way by the time that they are three weeks old.

▶ This beetle, which is found in deep caves, is completely blind. Its antennae are so sensitive that they can feel tiny movements of air, such as those made by another small creature nearby. The beetle is able to pinpoint and catch its prey.

WALRUS

Feeling Movement

The dark line that you can see down the side of the fish below is its lateral line system. It is the fish's most important sense organ. The lateral line is made up of a series of pits, each one sensitive to pressure. Any movement in the water, even when it is dark, will be detected by the lateral line system. With this system, a fish can find food and escape enemies.

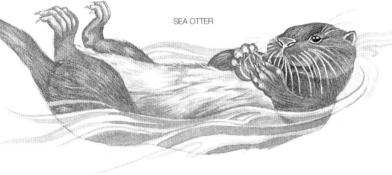

SEA OTTER

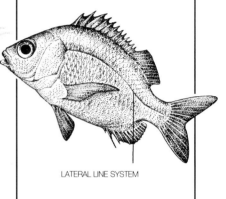

LATERAL LINE SYSTEM

SHREW

◀ The biggest and stiffest whiskers are around the faces of walruses, seals, and sea otters. These animals often dive into dark waters to hunt their food, and must use their sense of touch. Members of the cat family and shortsighted hunters, such as shrews, also have big whiskers.

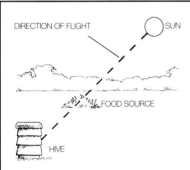

DIRECTION OF FLIGHT — SUN

FOOD SOURCE

HIVE

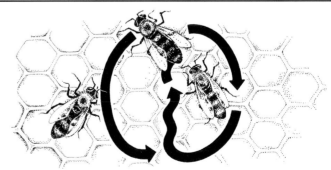

Dancing Bees

When a bee returns to the hive with food, she tells the other bees where to find it. The bee cannot give information by signals because the inside of the hive is dark. She does not use sounds because bees are deaf. Instead, she passes on her information by dancing. First she runs in a tight circle to get attention. Then she cuts across the circle, making a figure eight and waggling her **abdomen** as she does so. The other bees follow and touch her all the time. The direction of the figure eight tells them which way to go in relation to the position of the sun. The speed of her dance tells them how far to go. Heading up the comb means go toward the sun, as shown in the diagram above.

SMELLS

Smells, or scents, are caused by the chemicals that make up almost everything. We call the ability to detect scents the sense of smell. It was probably the first sense to have developed in animals many millions of years ago. Today it is still important to almost all creatures, but especially to animals that live in the dark. Scents can be used to give messages of many kinds, in water as well as on land.

An animal looking for food at night does not need to be able to see. It can discover food by its smell. Shrews and many other creatures that feed in semidarkness and night have long snouts for detecting smells. Moths feed on flowers such as honeysuckle, which give off a strong scent at night.

Groups of animals use scents as a way of keeping in contact and avoiding other kinds of creatures. Minnows can tell by smell alone at least 15 other kinds of fishes as well as different types of plants. The skin of some fish contains "alarm substances" which other fish can smell. If one fish is injured, the others in the school rush away as the alarm scent reaches them.

▼ Badgers, whose eyesight is poor, live in a world of smells. They have musk glands under their tails. They use their scent glands for many purposes. Badgers away from home, for instance, make scent trails so they can find their way back.

Snake's Tongue

All kinds of snakes, whether or not they are venomous, have forked tongues. The reason for this is that snakes, like some other hunting animals, track their prey by licking the ground. A snake flickers its tongue out and pulls it back into its mouth. In the roof of the mouth is a pit, lined with very sensitive smell and taste cells. These are able to detect the faint scent left by a mouse or some other prey animal as it passed by.

The snake's forked tongue enables it to hunt more effectively. It can cover a wider trail than a pointed tongue could.

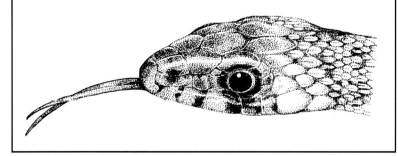

▲ Many kinds of animals defend themselves by using foul smells. The best known are the skunks, like the striped skunks shown here. Skunks are nocturnal hunters of small prey such as insects and mice. If they are attacked by a larger animal, skunks give a warning by flashing their black and white colors and drumming with their front feet. If this is not enough, they shoot a jet of foul-smelling liquid from under the tail onto the attacker. Any animal that has experienced this will never meddle with a skunk again.

▼ Here, you can see a coral reef at night. Coral polyps open their tentacles to catch **plankton** when it gets dark. A few fish, such as squirrel fish, hunt nocturnal shrimp. Sea urchins, feather stars, and basket stars also come out from hiding in the dark.

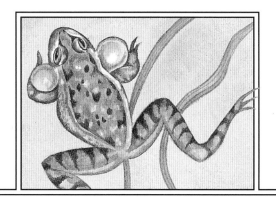

STAYING ALIVE

It is hard for an animal to stay alive in the total darkness of caves or the deep ocean. One of the main problems is that there are no plants. A few of the animals that live there eat dead plants and animals. Others are hunters. Finding a meal, but not being one, is not easy.

CATCHING A MEAL

In caves and in the deep ocean, meals are few and far between. If an animal is to survive, it must not miss a chance to eat. As a result, many of the animals that live in these two habitats are ferocious hunters, armed with teeth and jaws that do not allow prey to escape. In spite of this, almost all of them are small, because there is not enough food for large animals.

In the deep ocean there are fish with bodies that can stretch so that they are able to swallow a creature several times their own size. Cave animals tend to live far longer but to grow more slowly than their relatives that live in the light. This is probably because there is not much food.

▶ Hatchet fish, such as the one shown here, live in the open ocean at depths between 600 and 3,000 feet. Their chief food is small shrimplike creatures that form part of the plankton. At night they sometimes follow these to the surface of the ocean. This is a dangerous journey and hatchet fish, few of which measure more than 3 inches in length, are often caught and eaten by bigger species.

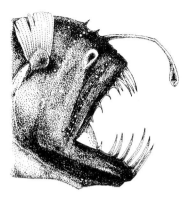

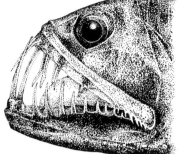

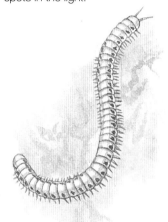

▼ Cave millipedes feed on dead plants washed into caves by streams. Like most cave animals, this millipede is blind and white. It has a line of poison glands down its sides, which show as bright red spots in the light.

Deep-sea Killers

Some of the most successful hunters in the deep ocean are angler fish. These have a fin ray that has become a fishing line, complete with bait that shines in the dark. The angler moves the bait back gently toward its mouth as a fish or shrimp is attracted by the light. When the prey is near enough, it is snapped up. Most deep-sea fish have a mouthful of needle-sharp, backward-pointing teeth. Some can dislocate their necks and push their delicate gills out of the way of struggling prey. This allows them to swallow a meal that is larger than themselves.

FINDING A MATE

Animals that live in the light have few problems finding mates. Males of many different types of animals use colorful displays to attract females and protect the place where their young are reared. This is not possible for nocturnal animals because they are not visible in the dark. Most nocturnal animals have drab colors and rely on sound or scent, or both, to find a mate.

Birds, which generally have little sense of smell, use their voices. Owls have what is known as an "advertising call," which tells other owls where they are. Many mammals, such as foxes, also use calls during the breeding season. Male frogs and toads sing loudly to attract females to the ponds in which they lay their eggs. These noises are often unmusical, but they can be heard very well in the darkness.

Other animals use scent signals. Mammals, especially those that live in dark forests, rely on scent more than on sight. But insects, including moths and many beetles, make the greatest use of their sense of smell. Females give off small amounts of scent. This is carried by the wind. Males detect it with their enormous antennae and follow the scent until they reach the female.

It is likely that cave animals, living in darkness all the time, also use scents to attract a mate. Unlike deep-sea creatures they do not make light signals.

FEMALE ANGLER FISH

PARASITIC MALE

▲ The difficulty of finding a mate in the vast oceans has been overcome by some of the angler fish. Large females carry several tiny males. The males are able to breathe for themselves, but do not need to feed, because their blood system is joined to that of the females. When the females release eggs, the males **fertilize** them. The eggs float up to become part of the plankton. As the young male fish begin to grow, they drift down to deeper water and find a mate. They remain with her for the rest of their lives.

Duetting Birds
These shrikes and some other kinds of birds that live in dark places have a special sort of song called duetting. The male sings a few notes and the female sings some others. These two songs are so perfectly timed, that together the sounds make a single song. Duetting is usually between birds that have already mated. It helps them to keep in contact, and perhaps to keep other birds away from their breeding area.

▶ A male *Callirhippis* beetle spreads his vast antennae when looking for a mate. These antennae are made up of a number of flat branches. Each one is covered with cells that detect the scent of a female. Other smells do not interest him. When the male is resting, he folds his antennae neatly under his legs.

MALE *CALLIRHIPPIS* BEETLE

◀ The spring songs of male frogs, like this Brazilian tree frog, range from birdlike trills to harsh croaks. Each kind of frog has its own call which is attractive to the female of that species. The song is loud and carries well through the night, because the frog's throat pouch acts as a sound box.

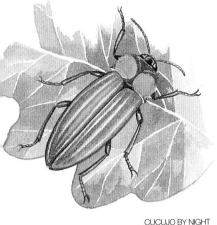

CUCUJO BY NIGHT

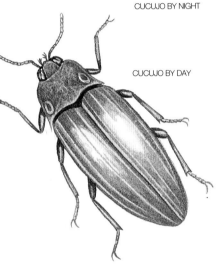

CUCUJO BY DAY

LIGHT AT NIGHT

Some animals deal with the darkness of the night or the deep ocean by making their own light. This living light is made by chemicals in the animals' cells. The basic chemicals are the same, whatever the animal. If you picked up a firefly, it would not burn you because the light is cold. The reason for this is that living light loses almost no energy as heat. Fireflies lose 2 percent of the energy that they use. An electric light bulb loses 97 percent of its energy in unwanted heat.

There are many kinds of land animals that make light. This light is often not very bright, and we do not know how it is used. On the other hand, some insects make very bright lights, which are used to attract mates. Male glowworms fly to the bright light made by the females. Both male and female fireflies light up. Fireflies often make brilliant displays, in which large numbers of insects flash their lights on and off at the same time. The brightest light is made by a South American beetle. This beetle has been used to light small rooms, or as a living flashlight in the forest.

◀ ▼ Here are some insects that make their own light. Whatever its color, living light is not changed by cool weather or rain, though some tropical fireflies do not display their lights when there is a full moon.

FIREFLIES

RAILWAY WORM BY DAY

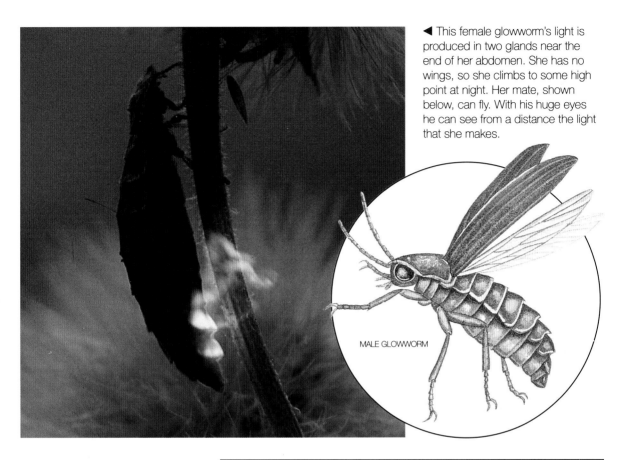

◀ This female glowworm's light is produced in two glands near the end of her abdomen. She has no wings, so she climbs to some high point at night. Her mate, shown below, can fly. With his huge eyes he can see from a distance the light that she makes.

MALE GLOWWORM

RAILWAY WORM BY NIGHT

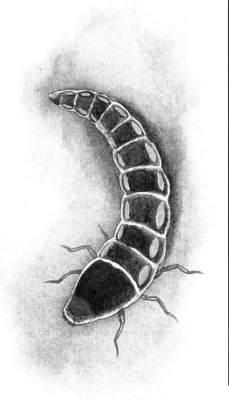

Gone Fishing

Huge numbers of **larvae** of midges live in the Waitomo Caves in New Zealand. These insects spin silk platforms, with long dangling "fishing lines" hanging from them. Each line is like a necklace with droplets of a sticky substance that glows in the dark. These droplets are put on the line, one at a time, by the larvae. If a small insect hiding in the cave is attracted to the beads of light, it gets stuck on them and is trapped. The larva then hauls the insect up to the platform and eats it.

LIGHT IN THE SEA

There are many kinds of ocean-dwelling animals that produce their own light. The light is sometimes caused by very tiny creatures called **bacteria**. One kind of fish keeps light-producing bacteria in two pockets. These pockets can be opened or closed, so that the fish can flash light when it needs to.

Most other creatures that produce light have special light-producing cells. Animals of the deep oceans more often have light organs – producing pink, green, or blue lights – on their undersides and near their eyes.

The light organs do many things. Probably one of the most important is to act as a signal between creatures of the same kind. When animals cannot see each other, the pattern and color of the light will enable them to keep together. This is very important at mating times. Lights are often used for attracting food and also as a way of startling prey or predators. Many fish and squids have a small light near their eyes so that they will not be blinded by an enemy. Some deep-sea squids defend themselves with ink that glows in the dark.

▶ The most brilliant living light at sea is seen in warm waters, where creatures such as Venus's girdle produce rainbow colors in the surface. The lights of deep-sea fish and squids are rarely seen, but these creatures come up at night, following the plankton near the surface of the water.

▼ In cool waters, usually in mid to late summer, it is sometimes possible to see the sea blazing with living light. This is almost certainly caused by very tiny organisms called *Noctiluca*. Each one of the many millions of *Noctiluca* in the water gives a flash of light as it is moved by the waves or currents. Together they light up the whole surface of the water.

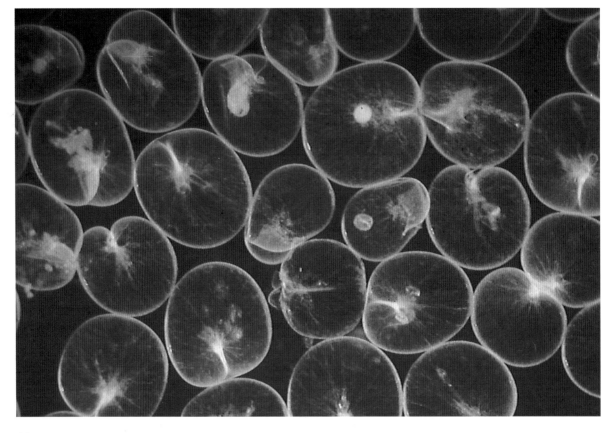

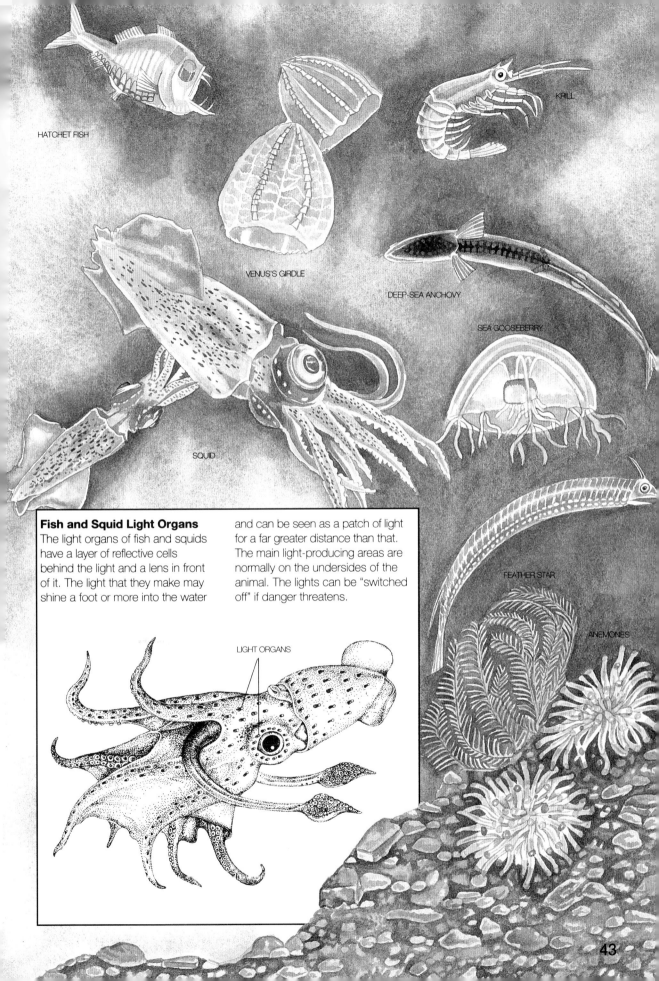

HATCHET FISH

KRILL

VENUS'S GIRDLE

DEEP-SEA ANCHOVY

SEA GOOSEBERRY

SQUID

FEATHER STAR

ANEMONES

Fish and Squid Light Organs

The light organs of fish and squids have a layer of reflective cells behind the light and a lens in front of it. The light that they make may shine a foot or more into the water and can be seen as a patch of light for a far greater distance than that. The main light-producing areas are normally on the undersides of the animal. The lights can be "switched off" if danger threatens.

LIGHT ORGANS

TRICKS IN THE DARK

To survive in the dark, animals must have senses that are very sharp. Some moths can hear the squeaks of hunting bats. Through eardrums on their sides, the moths can detect the sound of a cruising bat over 100 feet away, which is long before the bat is aware of the moth. As the bat approaches, the moth takes evasive action, turning and looping, spiraling and diving until it reaches the safety of some plants where it cannot be followed. Some moths can even make sounds that confuse the bat because they sound like the echoes of its own calls. They actually jam the bat's sonar. The bat's reply to these tricks is to make sounds that are too high for the moths to detect. Battles of this kind probably go on among many kinds of animals. We may not know about them yet, but we humans have still a great deal to discover about life in the dark.

GLOSSARY

ABDOMEN The hind part of an insect, or the belly of a vertebrate animal.

AMPHIBIAN A vertebrate animal. Generally the adults are land-living and breathe air, using small lungs. Some also breathe through their skin. The young are water-living tadpoles which breathe through gills.

ANTENNA (plural: ANTENNAE) The feelers on the heads of insects or other animals without backbones. They are involved in the sense of smell as well as touch.

BACTERIUM (plural: BACTERIA) Some of the smallest and simplest of all living things. The single cell that forms the body of a bacterium does not have a nucleus or other specialized parts, like the cell of, for example, an amoeba. Bacteria are found in almost all habitats, where they are of great importance, for they are recyclers, breaking down dead plants and animals and returning them to the environment.

CAMOUFLAGE A disguise that makes an animal difficult to see because it features shape and/or color that matches the animal's background.

CARNIVORE A flesh-eating animal.

CELLS Microscopically tiny units of which plants and animals are made.

COLD-BLOODED An animal whose body temperature is dependent on the warmth of its surroundings. On a cold day it will have a low temperature; on a warm day its temperature is far higher than that of a mammal. Because of this, it does not have a steady output of energy, but it needs very little food compared to warm-blooded creatures such as mammals and birds.

ECHO A sound bounced off an obstruction. The process of locating objects by means of sound waves is called **ECHOLOCATION**.

FOOD CHAIN The complex of plants and animals on which a creature depends for its food. It includes not only those that the creature eats, but those that affect the growth or survival of its food organisms.

FERTILIZE To bring together male and female sex cells so that a new generation is formed.

GLANDS Small parts of an animal's body that release chemical messengers into the bloodstream. The effect of these is to alter the animal's behavior in some way.

INNER EAR The inmost part of an animal's hearing apparatus that lies inside its head.

IRIS The colored part of a vertebrate's eye that surrounds the pupil.

LARVA (plural: LARVAE) The young of some animals. Larvae are able to fend for themselves, but they look different and live and feed differently from their parents. When fully-grown, they change fairly rapidly to the adult form.

MAMMAL A warm-blooded, air-breathing animal with a backbone, fed in the early stages of its life on milk produced by its mother.

NECTAR A sugary substance produced by flowers to attract pollinators.

NERVE ENDING The end of a nerve. It usually lies in or just below the skin. When stimulated it informs the brain of a change in, for example, pressure, light, and sound.

PARALYZE To affect the nervous system of an animal in some way so that it cannot move.

PIGMENT A chemical that gives color to animals, plants, or other objects.

PLANKTON Animals and plants that float in the currents in lakes or oceans. Most are small, many are the young stages of larger animals.

POLLINATE To transfer pollen from one flower to another.

PREDATOR A hunter.

PREY The animals caught and killed by predators.

PUPIL The dark area in the middle of the eye of a vertebrate, through which light passes. The pupil is often rounded in shape but may be oblong, or even zig-zag shaped.

REPTILE An air-breathing vertebrate animal, with a hard, dry skin, often armored with scales or bone. Reptiles are cold-blooded and their young usually hatch from eggs, though a few kinds give birth to live young.

RETINA The light-sensitive area at the back of an eye.

RODENT A mammal with two strong front teeth in its upper and lower jaws, which it uses for gnawing. Squirrels, mice, and porcupines are all rodents.

SENSES The special powers that animals have to discover things happening about them. We generally talk of five senses, sight, touch, hearing, smell, and taste. Some animals have extra senses besides these. Many animals have senses far better developed than ours. For instance, a dog has a sharper sense of smell, a bird has better eyesight.

SWIMBLADDER A gas-filled bladder inside a fish. This acts like a built-in lifebuoy, so that the fish is weightless in water. As a result, all of the energy used in the fish's movement goes to push it forward, rather than keep it up in the water.

TENTACLES Long, slender, arm-like parts of some kinds of animals. They are used for feeling and holding things, and sometimes for moving.

VENOM Poison produced by animals. It is usually injected into prey creatures to kill them, but it may be used defensively.

VIBRATE To move rapidly and continuously to and fro.

INDEX

Illustrations are indicated in **bold**

A TEMPLAR BOOK

Devised and produced by The Templar Company plc
Pippbrook Mill, London Road, Dorking, Surrey RH4 1JE
Copyright © 1991 by The Templar Company plc

PHOTOGRAPHIC CREDITS